Beginning Astronomy with a Celestron Equatorial Mount ...
So, what's the matter?

Jim Meadows

Practical solutions to questions when starting Astronomy with an Equatorial Mount*

*Based on using a Celestron® CG-5 mount. With other mounts your mileage may vary.

Copyright © 2013 Jim Meadows
All rights reserved.

Preface

This is my second book of practical answers to problems and questions when beginning a new area of Astronomy. My first book, "Beginning Astronomy with a Celestron ... So, what's the matter?", dealt with finding solutions to problems when starting Astronomy with a computerized telescope. Many questions dealt specifically with using the Celestron 8se on an altitude-azimuth GoTo mount, but some questions also applied to other telescopes as well.

I decided to step up to an equatorial mount to use with my 8se and purchased a Celestron CG-5 computerized German Equatorial Mount and ultimately was very pleased with the improvements of more accurate GoTo's, better tracking, etc. However, I ran into various issues while making the switch. There were a lot of extra steps and I was a little confused by the setup, align, polar align and realign process. I wondered how I should go about balancing it with counterweight(s). And how do I point this thing? I began to wonder if it was going to be too complicated and should I return it. And I thought it sure would be helpful if there was a practical guide to changing from an altitude-azimuth mount to an equatorial mount.

Like before, as I came up with solutions to the various problems I ran into, I wrote them down. I also found some solutions on the Internet. I have collected many of the common issues you may encounter when switching to an equatorial mount and turned them into book #2.

By the way, the Celestron CG-5 computerized GEM is a great mount to begin learning to use an equatorial mount. As this book was being published, Celestron released a new Advanced VX mount very similar to the CG-5 but with several improvements. So you may be able to get a CG-5 at a good price as the Advanced VX gets into the market. Much of the contents of this book should also apply to the Advanced VX if you decide to go with the latest greatest model.

As I said in my first book, I hope you find some suggestions here that make your life easier in this great hobby of Astronomy!

Jim Meadows
June, 2013

Disclaimer

The information in this book is presented as accurately as possible based on the information the author had at the time of publication. The author makes no guarantees regarding the information contained in this book. However, the author has worked hard to ensure that the information is as accurate as possible at the time of publication.

The author does not receive income from any of the products described in this book.

Acknowledgements

Celestron and NexStar are registered trademarks of Celestron, LLC.

ORION is a registered trademark of Orion Telescope and Binoculars.

Apple, Mac, iPhone and iPad are registered trademark of Apple Inc.

HyperTune is a registered trademark of Deep Space Projects.

All other proprietary names are the property of their respective companies.

Introduction

I am an engineer and like to solve problems. I also like Astronomy. Like I said in my first book, this is not a book about Astronomy though. There are many good books about beginning Astronomy and learning the night sky. This book is about common problems you may run into when starting Astronomy with an equatorial mount. It specifically addresses some items you may run into when setting up and using a Celestron CG-5 computerized German Equatorial Mount (GEM). The CG-5 is also known as the mount for Celestron's Advanced Series GT telescopes. Many items covered for the Celestron CG-5 also apply to other equatorial mounts as well.

This is not intended to be a replacement for your manual. Be sure to follow the manual as much as possible. This book contains helpful information that is not detailed in the manual that will hopefully save you some time and frustration.

The format of this book is simple. It contains a list of questions with a discussion of how to resolve the problem at hand. All questions assume a viewing location in the Northern hemisphere unless otherwise specified.

Beginning Astronomy with a Celestron Equatorial Mount…

1. **So, what is an equatorial mount and why is it different from an altitude-azimuth (alt-az) mount?**

    An equatorial mount enables a telescope to follow the movement of stars by moving at a fixed speed around a <u>single</u> axis parallel to the Earth's axis of rotation, while an altitude-azimuth mount requires moving the telescope along <u>two</u> axes at the same time to track objects in the sky.

    An altitude-azimuth mount moves the telescope up and down (altitude) as well as left and right (azimuth).

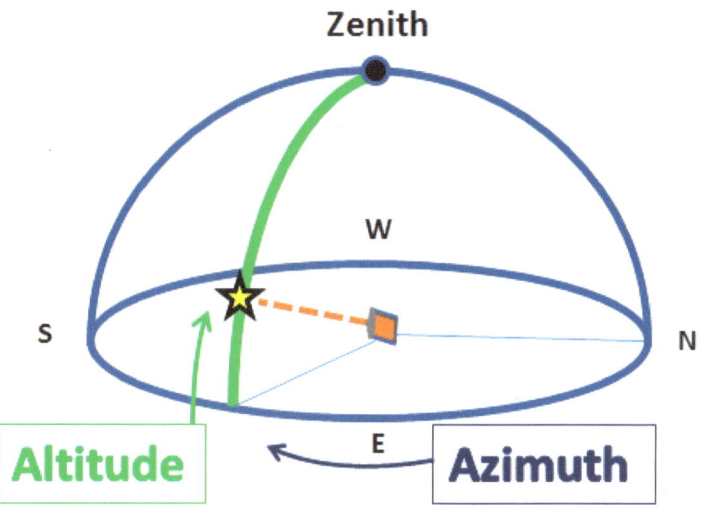

This makes it easy to point the telescope at a star or object, but it must continually be moved in both directions at varying speeds to stay centered on a star as the Earth rotates.

An equatorial telescope moves the telescope in a circular motion around the Earth's axis (Right Ascension - RA) and between the poles (Declination - Dec).

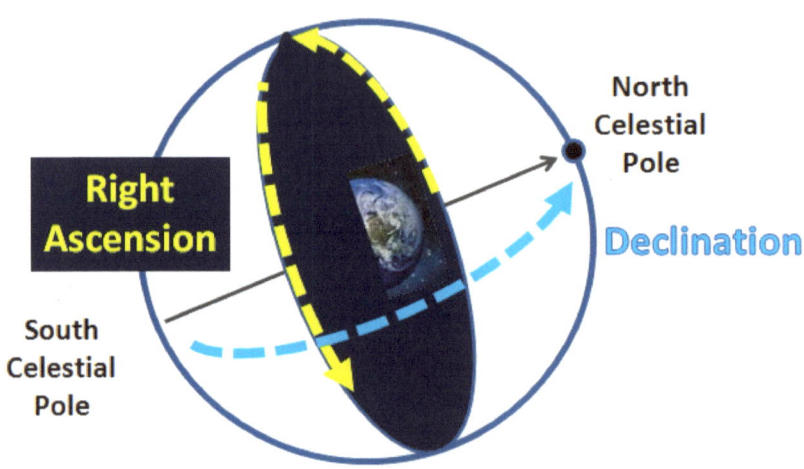

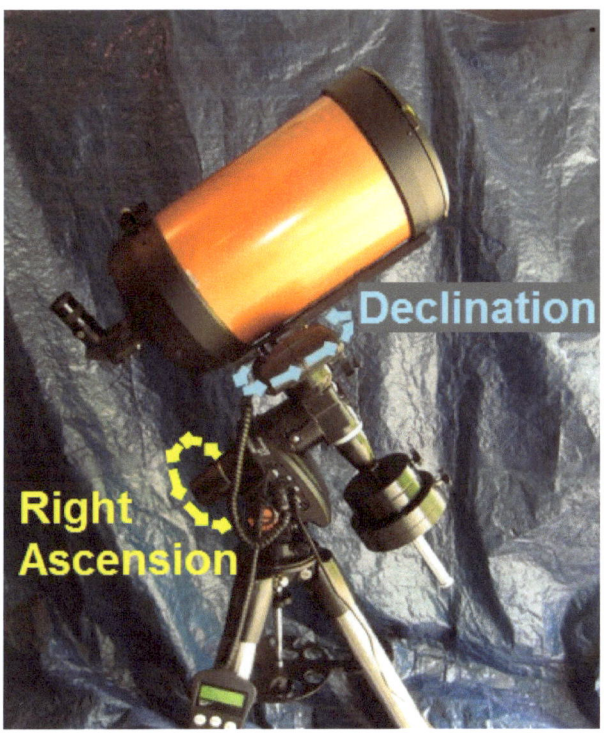

It must be properly aligned to the north (or south) celestial pole for it to work properly, but then only requires movement around the right ascension axis to track a star.

A motor, which is sometimes referred to as a "clock drive", can turn the telescope around the Right Ascension axis at exactly the speed needed to counteract the rotation of the Earth about its axis and keep objects centered in the eyepiece (one revolution every 23 hours and 56 minutes).

2. So, give me the Top 10 reasons to switch to an equatorial mount.

Here are some of the main benefits I have found switching from my Celestron alt-az mount to the CG-5 GEM.

#10 – It automatically slews to the stars for you during alignment.

#9 – In general, it is a lot easier to center objects in the eyepiece due to its balanced movement and less backlash issues.

#8 – It is typically better made and has more robust gear mechanisms for smoother tracking.

#7 – Once it is polar aligned, it tracks stars and objects really well and makes life a lot easier for astrophotography and video astronomy, with no field rotation.

#6 – It immediately starts approximately tracking the first star it slews to for alignment, allowing you to fine tune your focus and perform adjustments to center your finder with your telescope before you complete the first alignment step without the star drifting out of sight.

#5 – You can add calibration stars to increase the accuracy of your alignment.

#4 – Once aligned and calibration stars are added, GoTo's are more accurate and objects are positioned much closer to the center of your field of view.

#3 – It can point straight up without your eyepiece hitting the bottom of the mount.

#2 – You can put a lot more accessories onto your telescope since you use counter weights to balance everything evenly.

#1 - It really looks cool with its counter-weights and the way it moves around in circles using its gears and motors.

3. **So, give me the Top 5 reasons to not switch to an equatorial mount.**

And here is the other side of the coin…

#5 – Switching to an equatorial mount means you will spend some more of your hard earned money.

#4 – You have to set it up pointing north (or south) before you even begin.

#3 – The alignment process takes longer.

#2 – It is heavier and not as portable.

#1 – You have to take time to put counter weights on it and balance everything and it is harder to figure out how to point it. (Of course this gets offset by the #1 reason to switch to an Equatorial mount).

4. Is a German Equatorial Mount better than other mounts?

A German Equatorial Mount (GEM) does not mean it is made in Germany. It refers to a specific type of equatorial design that is shaped like a "T" and uses counterweights to balance the weight of the telescope. See the next question for more info on various types of equatorial mounts.

You can certainly get high quality alt-az mounts that will perform just as well as an equatorial mount. As with most anything, the more you pay for either type of mount, the better it will probably perform. My specific experience was well worth the move up from the alt-az mount that comes with the Celestron 8se to the CG-5 GEM. If you have more money available to spend, then you may want to consider the upgraded Advanced VX model or the even heavier duty Celestron CGEM model, or equatorial mounts available from others.

5. So, what are the different kinds of equatorial mounts?

The most common type is probably the German Equatorial Mount like the CG-5 mount. Below are some of the various types of equatorial mounts. All of these use some technique to rotate the telescope around an axis that counteracts the movement of the Earth around its axis. Search Wikipedia for Equatorial Mounts to see pictures of these types.

a. German Equatorial Mount – this uses counterweights along one axis opposite the telescope and is commonly used by amateur astronomers.

b. Open Fork Equatorial Mount - this resembles an altitude-azimuth Fork Mount but with the azimuth axis tilted and lined up to match the earth's rotation axis using what is typically called a wedge. This is another design that is used by amateur astronomers.

c. English or Yoke Equatorial Mount – this has a frame with the right ascension axis bearings at the top and bottom ends and the telescope inside the midpoint of the yoke. There are no counter weights and is mainly used for larger professional installations.

d. Cross-axis Mount – this mount is like a big "plus" sign with the right ascension axis supported at both ends and the telescope mounted at its midpoint on a declination axis with a counter weight.

e. Equatorial Platform Mount – this is really a platform that allows you to put anything you want on it to track on an equatorial axis from small cameras to entire observatory buildings.

6. So, can I convert my alt-az mount to an equatorial mount?

I actually tried converting my Celestron alt-az mount to an equatorial mount by adding Celestron's heavy duty wedge as shown below.

It is mounted onto the tripod to effectively tilt the azimuth axis of the alt-az mount so it can match the earth's rotation. However, I found that it put a lot of strain on the one arm design of the 8se's alt-az mount and you are limited on how much you can attach to the telescope.

If you have a fork alt-az mount, then it could be more practical to add a wedge so it can function as an equatorial mount since the telescope is supported on both sides. If you are moving up from the basic Celestron alt-az mount, my recommendation is to skip the wedge and put your money into a full equatorial mount like the CG-5. The smoother gears, better tracking and ability to add more accessories was well worth it for me.

7. So, could I just add an autoguider to my alt-az mount to get the same tracking accuracy as an equatorial mount?

This is another "it depends". I tried an autoguider on my 8se alt-az mount and did get better tracking. But it takes it own setup time to find a good guide star and requires recalibration as you move to different parts of the sky. And I still had issues with its gears and backlash to contend with. Adding an autoguider typically also means having a computer on hand to perform the autoguiding. I can do beginning astrophotography and video astronomy on the CG-5 without having to use an autoguider.

The other issue you run into with an alt-az mount is field rotation when performing long exposures, even with an autoguider. Field rotation is the apparent rotation of celestial objects around the center of the field of view of a telescope over a period of time. It doesn't really impact visual observing, but it cannot be ignored for astrophotography. All the stars will appear to move around a central point or star that is being tracked. Only the center will show pinpoint stars or sharp detail. Toward the edge of the field of view, all stars will show as concentric arcs around this center. Field rotation will occur unless the mount is aligned to counteract the earth's rotation, which is the basis of an equatorial mount, and cannot be done with an alt-az mount.

8. So, what is involved in setting up a GEM?

There are more steps to set up a GEM, but the payoff during your viewing time is worth it to me. However, be prepared to learn some additional setup techniques when switching to a GEM. Here is an outline of the general setup process I use for my CG-5.

   a. Set up the tripod pointing North
   b. Center the mount onto the tripod with the azimuth housing over the metal peg and hand tighten the mount into place.
   c. Attach the eyepiece tray.
   d. Level the tripod and mount by adjusting the legs.
   e. Turn the altitude adjustment screw to your approximate latitude.
   f. If you are outside at night, look through the polar axis of the mount and move your tripod to approximately center the mount on Polaris.
   g. Attach the hand controller and the declination cord.
   h. Add the counterweight(s) onto the mount.
   i. Attach the telescope to the mount.
   j. Add your eyepiece and other accessories you plan to use.
   k. Adjust the counterweight(s) to balance the telescope.

1. Turn on the mount and begin your alignment process.

Note that you should set up the tripod pointing north and center Polaris through the polar axis of the mount before you attach the counterweight(s) and telescope. It is much easier to move it without the additional weight in place.

9. So, how much time should I spend leveling my tripod?

Don't spend a lot of time leveling your computerized mount. It mainly matters for two things – keeping the mount from tipping over and slewing to the first couple of stars in the alignment process. If your mount is not completely level, it only changes your apparent position which will be corrected during the alignment process.

For an equatorial mount to function properly, the main requirement is for the mount's Right Ascension axis to be parallel to Earth's rotational axis. So as long as you can use your adjustment screws to point the RA axis toward the celestial pole, then your mount will be properly aimed regardless of how level the tripod is. Then, after you have centered it on two alignment stars, the computer will compute its actual position. If you perform an All Star Polar Align it will be even more accurately positioned, regardless of how level the tripod is (it does make it a little easier to perform this process when level).

10. So, how tight should I screw the tripod rod into the mount?

When you push up on the rod under the tripod to attach the mount, just hand tighten the tripod rod knob to secure the mount to the tripod.

Do not over tighten since the mount will need to be able to be rotated slightly on the tripod using the azimuth screws to polar align the mount. If you tighten it too much to the tripod, you may place excess stress on the metal peg when adjusting the azimuth screws and cause the peg to loosen. See question 27. If you find it hard to adjust the azimuth screws, you may need to loosen this rod a little.

11. So, how much should I screw in the knob to tighten the eyepiece tray?

Again, hand tighten the knob when attaching the eyepiece tray enough to ensure the legs are fully spread out, but do not over tighten for the same reason noted in the previous question.

12. So, any suggestions when balancing GEM?

Make sure you have all your accessories attached that you generally plan to use so you will be balancing it as it will be used. Go ahead and attach your lens shade/dew shield if you are going to use one since it definitely affects the forward/backward balance of the telescope.

When you slide the counterweight onto the bar, make sure the larger hole is on the side facing down. This is so you can move the counterweight all the way down over the safety screw at the bottom for balancing if needed. Put your counterweight(s) on before attaching the telescope tube. Then, be sure to have the declination and right ascension clutch knobs tight when you attach the telescope and anytime you slide the telescope forward or backward on the mount.

Balance the right ascension axis first as described in the manual to adjust your counterweight(s). You will note in the instructions that you may want a slight imbalance, depending upon the direction the telescope will be pointing if you plan on viewing a specific area of the sky for a period of time. For general use I just try to get it evenly balanced.

It is also a good idea to then balance the declination axis as described in the manual by shifting the telescope forward/backward as needed due to the accessories you have attached. When moving the telescope forward/backward, I always return the RA axis back to its index marks and hold the visual back of the telescope with one hand while I adjust the telescope's position on the mount. I then retighten and recheck to see if this balanced it properly.

13. So, why are there two screws to tighten the telescope onto the mount?

The two screws ensure the telescope will not slip during use by having two locking points. Loosen both screws before slipping the telescope into the slot on the mount and then snug them both up.

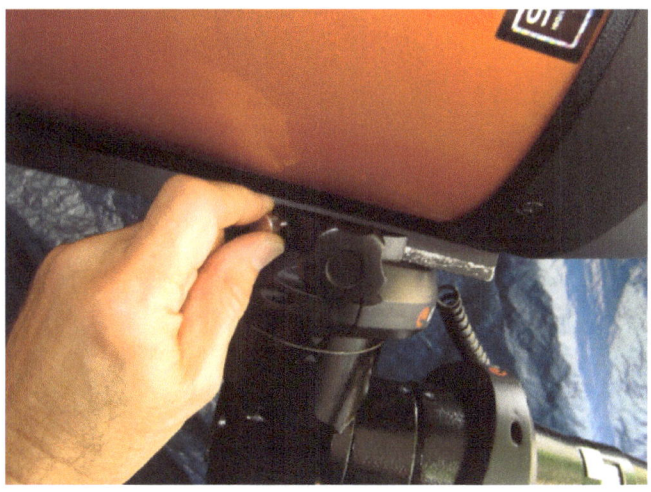

Later when you balance your telescope, place one hand on the telescope and loosen the small safety screw. Then loosen the main knob until you can move the telescope forward/backward to place it into a better balanced position and tighten it back up. Then retighten the small safety screw.

14. So, will I need additional counterweights and what size?

My CG-5 came with one 11 pound counterweight. As I began to use more accessories (cameras, etc) I needed more weight to balance it, so I purchased another 11 pound counterweight. It seemed though that generally this was a lot more weight than I needed. I found out that an Orion AstroView 7.5 pound counterweight is compatible with the CG-5 and purchased it, and that is what I use primarily as my second weight. It also enabled me to balance some smaller telescopes on the CG-5 as well.

15. So, I never used the setting circles markings on my alt-az mount, do I need to use them on an equatorial mount?

The short answer is "not really", particularly if you have a computerized mount. Setting circles are used to manually position the telescope to an approximate RA hour and Dec degrees location in the sky. If you use a GoTo mount to slew your telescope to objects in the sky, you will not use these markings.

16. So, what are those circles with numbers under the removable cap for?

The set of markings underneath the removable polar finder cap can be used during a manual polar alignment. Once calibrated for your specific location, you can use them to rotate your RA axis based on your current time in order to properly position your mount using Polaris. A more detail description of this can be found at www.nightskyimages.co.uk/polar_alignment.htm.

However, if you use a polar scope that includes the major constellations around Polaris, you can just rotate the RA axis until you align the constellations etched inside the polar scope to match their position in the sky, and then adjust your mount until Polaris is in the indicated position.

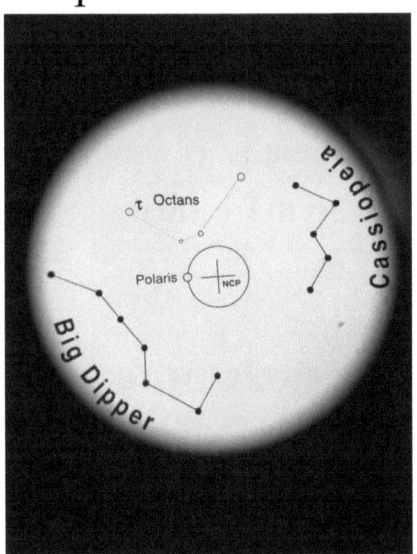

So, if you install a polar scope with constellations you will probably not use the time notation markings at all.

17. So, why does the equatorial mount alignment process take longer?

By design an equatorial mount must be pointed at the north Celestial Pole (or south if you are in the southern hemisphere) in order to use one of its axes to counteract the motion of the Earth around its axis. So this means you have to take some time to get it pointed properly.

But it also means it will really pay off during your viewing session with better tracking and more accurate GoTo's.

There are some minimum steps you can take, but as you spend time adding additional steps your alignment will be even more precise. What you plan to do during your viewing session will determine how much time you take to align your equatorial mount.

18. So, what is done for basic alignment?

At a minimum you will need to do the following alignments:

a. Set up your telescope. See question 8. This will include manually pointing your tripod North and turning the altitude adjustment screw to your Latitude

b. Turn on your mount and press Enter. You will be prompted to move the telescope mount so the index marks for both RA and Dec are aligned, and to enter your current time and location (city or latitude). You can either loosen the clutches and manually move the mount to align the index marks, or press the arrow keys on the hand controller until the index marks line up.

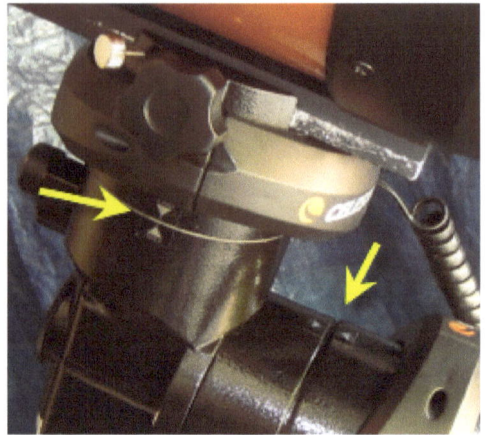

c. Select a Two Star Align (or a One Star / Solar System Align).
d. Follow the prompts to center the alignment stars. When prompted to add calibration stars, you can press the Undo button to exit.

If you are in a hurry and plan to look at bright objects like the moon and planets, you can select quick align in step c which skips having to center any alignment stars. This allows you to slew to objects and basically track them. You can add alignment stars and calibration stars later to improve GoTo's and tracking accuracy.

19. So, what additional steps can be used for better alignment?

A better alignment process would include:

a. Set up your telescope. See question 8. This will include manually pointing your tripod North and turning the altitude adjustment screw to your Latitude

b. Look through the polar axis of your mount to better center Polaris. You will need to remove the two polar axis caps and rotate your declination axis perpendicular to the polar axis to be able to see through the mount's polar axis. Then use the altitude and azimuth adjustments screws to center Polaris.

c. Turn on your mount and press Enter. You will be prompted to move the telescope mount so the index marks in RA and Dec are aligned, and to enter your current time and location (city or latitude).

d. Select Two Star Align.

e. Follow the prompts to center the alignment stars.

f. Add one or more calibration stars when prompted. You can press undo when it prompts for the next calibration star if you use less than four.

Note that for step b, if you have a polar scope installed in your mount, you will move the mount to get Polaris to the indicated position seen through the polar scope as shown in question 16. This will improve your initial GoTo slews during alignment as well. You may notice the latitude setting on the side of your mount has changed after you perform step b. This does not matter though, as long as your RA axis is now properly aimed toward the celestial pole. See question 9.

Using an eyepiece with an illuminated crosshair reticle really helps to accurately center a star. If you finish centering the object using the Up and Right direction buttons on the hand control, it will eliminate the effect of mechanical backlash from the gears and give you a better alignment.

20. So, what optional steps can be added for even better GoTo's and tracking?

If you are going to be doing any viewing that requires better GoTo's or tracking (e.g. deep sky objects), you probably should take the time to perform a full polar alignment as follows:

   a. Set up your telescope. See question 8. This will include manually pointing your tripod North and turning the altitude adjustment screw to your Latitude
   b. Look through the polar axis of your mount and use the altitude and azimuth adjustments screws to better center Polaris (or position Polaris properly if you have a polar scope installed).
   c. Turn on your mount and press Enter. You will be prompted to move the telescope mount so the index marks in RA and Dec are aligned, and to enter your current time and location (city or latitude).
   d. Select and perform a Two Star Align. Press Undo to skip adding calibration stars at this point (or just add 1 and then exit).

e. Perform an All-Star Polar Align (ASPA) to better position your mount to the true celestial pole by pressing the Align button and selecting Polar Align / Align Mount and follow the prompts. In this process you will initially center a star using the hand controller, and then center it again when prompted by manually adjusting the mount's direction using the altitude and azimuth adjusting screws.
f. Turn off your mount when finished with the ASPA.
g. Turn on your mount repeating steps c and d to perform a basic two star align and add up to four alignment stars.

When you select the All-Star Polar Align, it will use the last star you slewed the telescope to. It does not use Polaris for the star, thus the "All-Star" name. So, if you have just performed a two star align, it will use the last star you centered during alignment. You may want to use a different star though. For best ASPA results, you should use a star that is somewhat high in the sky and near the Meridian (the east/west crossover point), but not directly overhead. If you see a better candidate than your last alignment star, use the hand controller to select that star and slew to it before doing the ASPA in step e. If you don't know the star's name, use the arrow buttons to slew to it (see question 30), press the Identify button to find its name and re-slew to it.

If you are going to perform long exposure photography, you may want to repeat the All-Star Polar Align process a second time (steps c-f) before proceeding to the final alignment (step g). Really long exposure photography can get into even more sophisticated additional Polar Alignment procedures (e.g. drift alignment). See question 21.

There are two functions under Calibrate Mount in the Utilities menu that can also improve performance. One function calibrates the RA index marks, which will help the mount more accurately locate your alignment stars. Use this one time after a good alignment and it will save the index correction values. The GoTo Calibration function should be used whenever you attach heavy accessories to your telescope to optimize the time it takes to slew to objects.

As I said, the amount of time you take to align an equatorial mount really depends upon how you plan to use it for the evening. You don't need to do All-Star Polar Align's if you are just doing basic visual observations. So adjust the align process and time to match your needs.

Another way to improve your GoTo's is to press Menu and select Precise GoTo to enable the hand controller to better center those faint Deep Sky Objects for you. It will first automatically slew to a bright star close to the object you selected and prompt you to center the star in the eyepiece. It will then slew to your object with enhanced accuracy. If you are going to be observing several objects in the same region of the sky, using the Align / Sync function on a bright object in that area will make all your GoTo's in that area more precise. Don't forget to select Undo Sync when moving to a different region of the sky.

21. So, what are the different methods available to perform a polar alignment beyond just pointing the mount North?

Polar alignment is all about trying to get your mount's polar axis pointed at the north (or south) Celestial Pole as best you can to make maximum use of the equatorial mount's design. It is <u>not</u> about adjusting the telescopes position in RA or DEC using the hand controller. You are aligning your mount, not the telescope. All polar alignment procedures result in you physically adjusting the mount in altitude using the altitude up/down adjustment screw and adjusting the mount in azimuth using the azimuth left/right adjustment screws.

There are many ways to go about this process. Below is a list of methods in varying complexity that I have run across.

a. Simple alignment on Polaris: Center Polaris while looking through the polar alignment axis.

b. More precise alignment on Polaris: Look through a polar finder to adjust your mount until Polaris is at the indicated position.

For a short document describing this process see www.celestron.com/c3/images/files/downloads/1223669429_polaraxixfinder.pdf.

c. All-Star Polar Align: After performing a two star alignment press Undo, then Align / Polar Align / Align Mount. You will be guided to center a star first using the hand controller, and then center it again by manually adjusting the mount's direction using the altitude and azimuth adjusting screws while looking through the eyepiece.

d. Drift alignment: View stars though the eyepiece over a period of time to determine how much a star drifts and adjust your mount's position to reduce the drift. You choose one star due south near the meridian and make appropriate adjustments using the azimuth adjustment screws to eliminate east/west drift. You choose a second star near the eastern horizon and make appropriate adjustments using the altitude adjustment screw to eliminate north/south drift. For a description of this process see starizona.com/acb/ccd/settinguppolar.aspx, or take a look at andysshotglass.com/DriftAlignment.html for a good video of this process. Drift alignment takes a long time, typically up to 30 minutes. The longer you take, the better the alignment.

e. Software assisted drift alignment: Use software to assist and speed up your drift alignment process. See eqalign.net/e_eqalign.html, or wcs.ruthner.at/index-en.php or www.ccdware.com/products/pempro for packages where the software watches a star through a camera and steps you through how to adjust your alignment based on its analysis.

f. Software assisted precise alignment: Perform precise alignment using software to move your telescope via ASCOM to slew to two stars and calculate the offsets you need, and then follow prompts to adjust the altitude and azimuth adjustment screws. This process can be used with or without a camera. See www.alignmaster.de for more information.

g. Image assisted drift alignment: Use a CCD imager or DSLR that can take a 2 min time exposure to capture the drift as an image that will assist you in adjusting your mount in altitude and azimuth to eliminate the drift. See www.observatory.digital-sf.com/Polar_Alignment_CCDv1-1.pdf for a short document explaining the process.

h. Iterative alignment: Last but not least is this technique where you alternate between a star and Polaris for about 5-10 minutes. Perform a sync on your last alignment star and then have it slew to Polaris. Physically adjust the mount to remove about half the error between where it points and Polaris. Then have it slew back to the same alignment star, center and sync and repeat going back to Polaris to manually make another adjustment. Repeat a few times and you ready. See www.covingtoninnovations.com/astro/iterating.pdf for a further description of this.

22. So, do you really have to do a full polar alignment?

Here is another one of those" well, it depends". You will at least do a basic alignment as described in question 18. You can then press Align and scroll to Polar Align and then select Display Align to view your current alignment accuracy. It will show Azm and Alt accuracy in degrees, minutes and seconds. If they are both 15 minutes or less (the middle number) you will probably be fine for visual observing and no further polar alignment is required. You may want to press Align and select Calibration Stars to add more calibration stars for better accuracy.

If you have performed a good All-Star Polar Align on a prior night and have a method of marking your location of your tripod's legs so you can set it back up again with the feet in the same locations (see question 39), you may find you do not have to do the ASPA, especially if you left your mount on the tripod (even if you had removed your telescope).

23. **So, once you do a polar alignment why do you have to do another standard alignment?**

You have just spent your valuable time performing a really good All Star Polar Align. When you select Display Align it even shows your alignment is spot on. So why in the world should you turn off your mount, lose your calibration and have to perform another standard align?

The answer is found in the fact that you physically moved the position of your mount when you performed the All Star Polar Align. Yes, you do have your mount better pointed at the celestial pole. But when you moved your mount during the ASPA, all the initial alignment calculations in the hand controller are no longer valid. It no longer has an accurate model of the sky where you mount is now pointed.

So even though it sounds painful, when you finish manually moving your mount to complete a polar alignment, go ahead and turn off your mount. Position your mount back to the index marks, turn your mount back on and do another two star align and add 1-4 calibration stars. You will be glad you did.

24. So, why are calibration stars needed?

Your alt-az mount only used alignment stars and did not prompt you to add additional calibration stars. Now your new GEM is asking you if you want to add calibration stars. What is going on? Adding calibration stars is necessary to calculate and compensate for "cone" error inherent in all german equatorial mounts. Cone error is the inaccuracy that results from the optical tube not being exactly perpendicular to the mount's declination axis. The hand controller is able to determine the cone error value by using calibration stars on the opposite side of the Meridian from the alignment stars. The more the merrier.

25. So, how do you mount a polar axis finderscope?

Where does a polar finderscope go? It will be inserted into the Polar Axis of your mount (a polar finderscope does not attach to your telescope like other finder scopes.). Locate and remove the polar axis cover and cap.

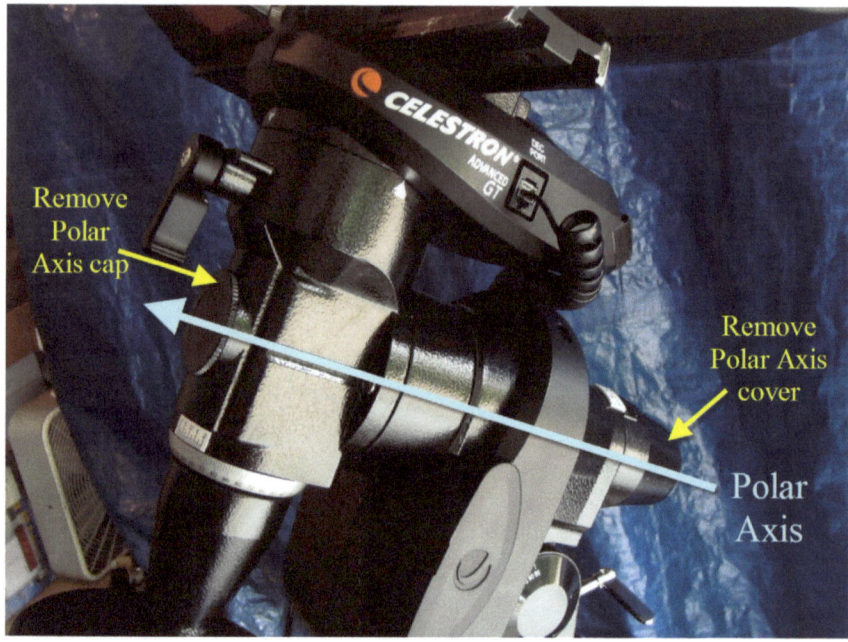

If you are going to install the CG-5 polar axis finderscope, you will need to remove a polar holder if there is one attached (it is for an older style polar axis finderscope). You need to unscrew it counterclockwise to remove it and then set it aside.

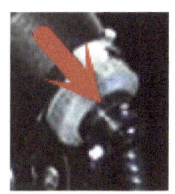 

Now insert the CG-5 Polar Axis Finderscope #94224 and thread it clockwise until tight.

When you use it, you will need to release the declination axis and rotate it perpendicular to the polar axis so you will be able to view unobstructed through the polar axis of your mount.

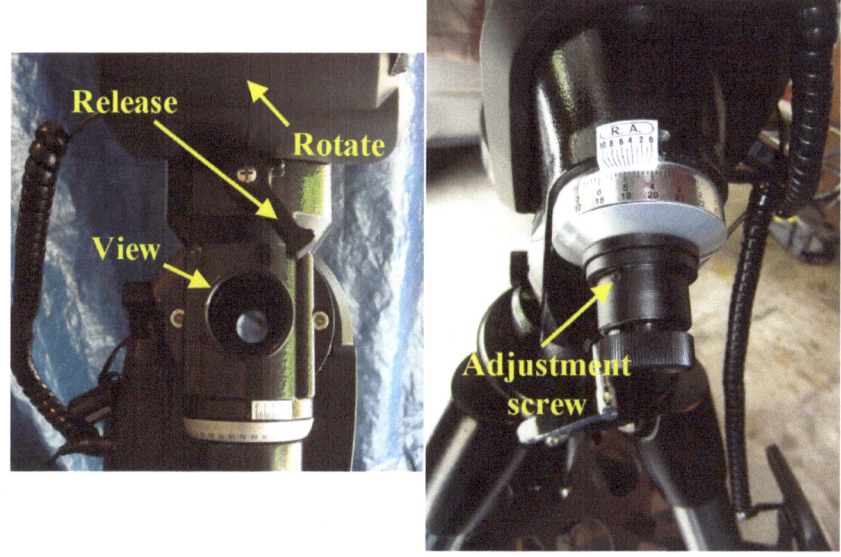

Follow the instructions that come with the polar finderscope to align its optical axis using the adjustment screens while looking up through the axis at an object (doing this during the daytime makes it a lot easier).

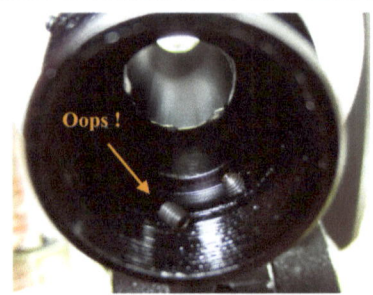

Warning: when turning the small adjustment screws only move them about 1/8 of a turn at a time while aligning the crosshairs. If you screw them in too much they can fall off inside the finderscope! Then you get to figure out how to take it apart to retrieve them by unscrewing the focuser and then put it back together again.

Since this finderscope does come with the constellation patterns etched in the reticle as shown in question 16, you do not have to use the time setting marks above the finderscope. In fact, I removed the small screw next to the RA label to eliminate any slight friction from the setting wheel.

26. So, how do I use the polar axis finderscope?

While using the polar axis finderscope, I store the cover upside down in the eyepiece tray and place the cap inside it to make it easy to find in the dark. Release the Dec axis and move it until you have an unobstructed view through the axis as shown in question 25 and retighten it.

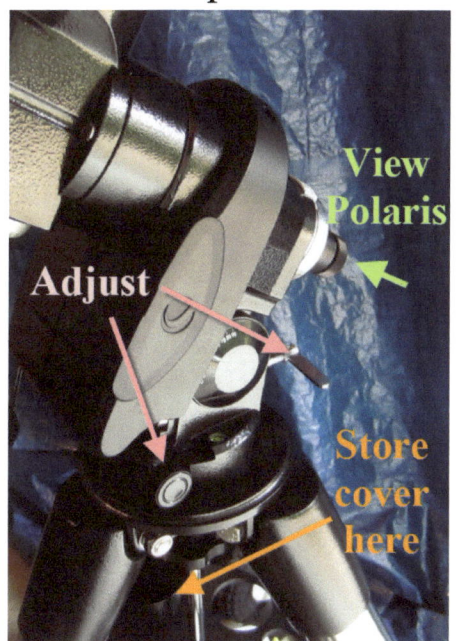

Now look up through the finderscope to view Polaris and the major constellations. Release the RA axis and move it until the Big Dipper or Cassiopeia match how they are located in the sky above you (you will not see them through the finderscope - you watch their etched images in the finderscope to get the axis positioned properly) and then retighten the RA axis. You may have to shine a dim red light down through the top end of the polar axis to better see the images and where Polaris should be positioned. Adjust the altitude screw lever and the azimuth screw knobs to move the mount so Polaris is in its indicated position. When done, be sure to turn both azimuth knobs until they are snug against the post under the mount. Once you have finished adjusting your mount, put the cover and cap back on the Polar Axis ends.

27. So, why does my view jump when I am adjusting the azimuth knobs during a polar alignment?

This means the metal peg on the top of the tripod that the azimuth screws push against has become loose. I believe this happened to me because I tended to tighten the rod mount knob too tight while attaching the mount, and then really tightened the eyepiece tray against the legs. I wanted to make sure the legs were not going to move.

Well, that means you have to apply a lot more pressure on the azimuth screws and thus on the metal peg they press against in order to move the mount left/right during a polar alignment.

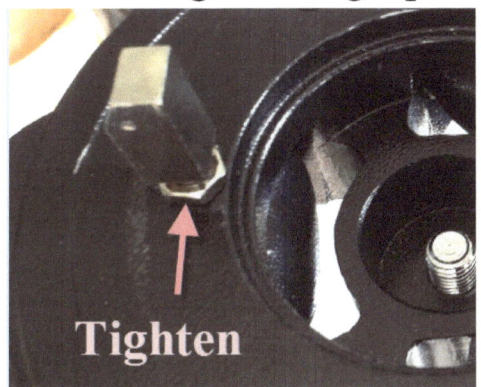

I finally realized my metal peg was starting to turn a little as I tried to adjust the azimuth knobs. So I used two wrenches to straighten up and tighten down the metal peg on top of the tripod. Then I started making sure the mount rod knob and eyepiece tray knobs were snug, but not too tight. Adjusting the mount using the azimuth knobs is now a lot smoother with no jerks.

28. So, what sighting / finder scopes should I use?

I highly recommend you go to your computer right now and order a Telrad reflex sight! It is an expensive but very effective tool to assist you during your initial alignment. It is also very helpful when manually pointing an equatorial mount. It is referred to as a unity finder (1x) since there is no magnification - you see a generous portion of the sky as you look through it. It has red rings that outline areas typically covered by finderscopes. The brightness of these rings is easily adjusted using a lever to select any level of brightness. It is extremely lightweight and removable from its base (the base attaches to the telescope with double sided tape).

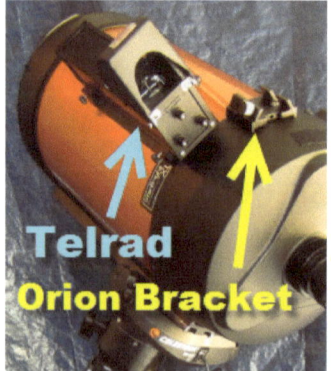

There are other products that are similar and smaller that can also be effective, but I found the Telrad worked better for me than a red dot sighting scope that only has fixed click stop levels of brightness. Be sure to also get the Telrad dew cover accessory that includes a mirror. This allows you to cover it when not in use to keep dew off, and the mirror can be adjusted for right angle viewing if your scope is pointed near vertical. Due to the accuracy of the equatorial mount once aligned, I have found that often I don't even use a finderscope.

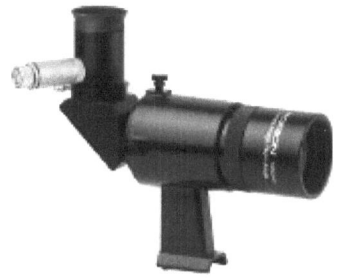

For when I do use a finderscope, I upgraded from the one supplied with the Celestron 8se to an Orion 9x50 illuminated Right-Angle Finder Scope that comes with a quick release mounting bracket for easy removal. The adjustable illuminated double crosshairs really assist precise aiming, and the right angle makes it easier to use. I am sure other similar finder scopes would do the same. The quick release mounting bracket can also be used for other accessories such as a solar finder or guidescope. See questions 29 and 47.

29. So, is there a special solar finderscope for pointing your telescope at the sun?

I need to start this answer with a disclaimer like you see in all the manuals. NEVER look at the sun directly through the eyepiece of a telescope without a solar filter in place over the end of the telescope or else you could permanently damage your eye. See question 1 for a photo of my solar filter in place. Of course this warning applies to finderscopes as well.

You can purchase from Orion a 2.32" ID E-Series Safety Film Solar Filter that will fit over the Orion 9x50 finderscope described in question 28 so you can use it to aim your telescope at the sun. Note that to even be able to select the Sun when using Solar System Align you must first go to the Utilities menu and select Sun Menu then press Enter in order to allow the Sun to appear as an object. You may also want to select Tracking Rate and change tracking from Sidereal to Solar for better tracking of the Sun. Just be sure to change it back when done.

If you cannot find a solar filter for the finderscope you use, a way that is often described is to watch the shadow of your telescope until it casts a minimum shadow indicating it is pointed at the sun. I have never been very good at this, though, particularly with an equatorial mount.

One inexpensive handy accessory that makes it easy to point the telescope at the sun is the Tele Vue Sol-Searcher Solar Finder. Just velcro it to the top of the telescope tube, or if you have already mounted an Orion quick release bracket, you can get the Tele Vue QRB-1006 Dovetail, attach it to the Sol-Searcher and slip it into the Orion bracket.

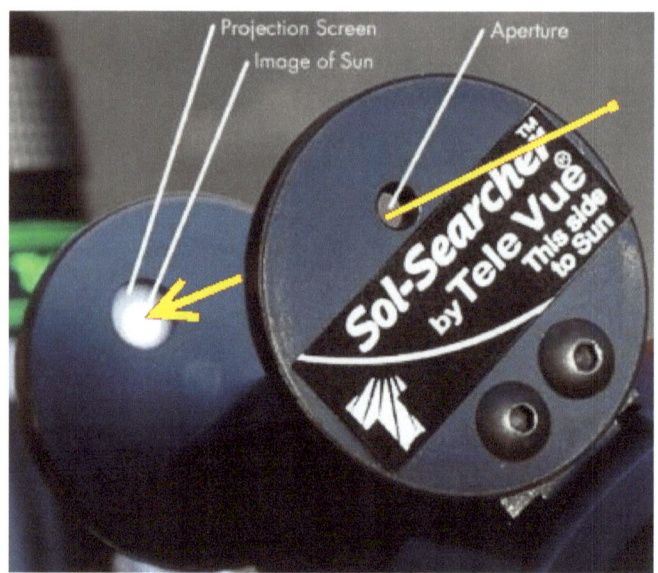

Slew the telescope to the approximate direction of the sun and watch the back portion of this little accessory as the light of the sun passes through the small opening at the front. Move the telescope until the small image of the sun is projected onto the back white circle and then you should be able to see the sun through the eyepiece of your telescope for focusing and viewing.

30. So, how do I aim this thing?

The simple answer is whenever possible let the hand controller computer do it! But that doesn't help when you need to do it yourself.

Unlike an altitude-azimuth mount which moves the telescope up/down and left/right, an equatorial mount's movements are tailored to the earth's axis and tracking the circular motion of the stars, and thus aiming it is not intuitive. This makes pointing an equatorial mount yourself an interesting exercise.

If you have the mount turned off, you can release both axis and move the telescope by hand to get it into a specific viewing position. In fact, it is a good idea to move it around some during the day to just get a feel for how an equatorial mount moves. You will begin to see how it is best to have the counterweight on the west side if you are pointing the telescope to the west portion of your sky, or on the east side when pointing to an object in the eastern sky. You can also see what happens if you try to track an object as it moves through the east/west point above you. You will see how you will need to flip the telescope around so that the counterweight is now on the other side. This is often called the meridian flip issue. See question 31.

When the axis clutches are engaged and you are using a computerized hand control to manually aim the telescope at an object, pointing it becomes a little more difficult since you typically move it along one axis first and then the other axis using the buttons on the controller. Here are some guidelines I use to assist me. As stated above, if an object is in the eastern sky, you will want the counterweight on the east side of the mount. Or if you need to point to the western sky, the counterweight will need to be on the west side of the mount. So the first thing I do is set the rate to 9 and press the right or left arrow button to start moving the RA axis in the direction that will have the counterweight on the proper side. You may be able to do this in a single movement. However, if you are switching from east to west or from west to east, you may start noticing the telescope will begin to point down at the cross over point. If so, you should stop and press the up or down arrow button to move the telescope back to an upward position by rotating it around the Dec axis. Then resume moving it around the RA axis to its approximate position.   Now you can move it around the Dec axis until the telescope is pointing in the general direction of the object.

At this point, I find using the Tetrad (or any red dot finder) a great help when making the final positioning movements. Unlike the standard up/down/left/right of an alt-az mount, the buttons move the telescope differently depending upon where it is pointed in the sky (which side of the Meridian the telescope tube is on). If you use a unity finder, you are looking directly at the sky with illuminated circles (or a dot) and can quickly get a feel for how each button is moving the telescope. Change the rate to a value of 6 or lower for the final movements.

Here is an example where the telescope is at the index marks and you want to point it to an object in the east as you are looking south.

Move it around the RA axis by pressing the right arrow button until the counterweight is on the east side. Then rotate the telescope around the Dec axis using the down arrow button to point the telescope to the east. Look through your Telrad (or equivalent) as you fine tune your final pointing.

Now let's say you decide you want to move it to an object in the west, so you start moving it around the RA axis to get the counterweight on the west side of the mount by pressing the left arrow button. As the telescope passes the east/west point, you realize the telescope starts to point downward so you stop.

You need to get the telescope rotated around the Dec axis so you press the up arrow until it points in the opposite direction. Now you can press the left arrow button again to resume moving it around the RA axis to finish pointing it in the general westward direction needed. As before, look through your Telrad (or equivalent) as you fine tune your final pointing.

You can adjust the hand controller settings to change the direction the buttons move the telescope if you want. Press the Menu button on the Celestron hand controller and select Direction Buttons from the Utilities menu to change the setting from positive to negative to reverse them.

Instead of changing settings, I usually just rotate the hand controller in my hands so that the arrow buttons "match" the direction I see a star move in the eyepiece and then tell myself, OK, at the moment I'm pushing or pulling the stars when I press this button (e.g. pushing it up/down and pulling it left/right).

31. So, what is meridian flip and can it break anything?

Meridian flip is an issue found with German Equatorial Mounts due to the need to keep the counterweight on the proper side of the mount. See question 30. It occurs when tracking a star or object as it moves through the east/west point above. While the object is in the east, the counterweight will be on the east side of the mount. When the object passes to the west side, as it continues tracking, the telescope starts tilting further down and eventually the assembly would come into contact with the motor housing. This may not break anything, but it is not a good thing to happen. So the mount has RA limits to keep it from tracking too far past the meridian, which means your object will drift out of the view from the eyepiece when it gets to the limit.

    If you start viewing something in the east close to the Meridian you will need to monitor this. At some point you will need to reposition the telescope so the counterweight is on the other side of the mount to keep viewing the object. An easy way to do this is as follows, once you know it has passed the Meridian, just select the object again and the telescope will reposition itself with the counterweight on the west side. You will notice the image is inverted after you reposition the mount and telescope. If you plan to view an object that is close to the Meridian, you may want to wait to select it until it has just passed to the west side of the Meridian so you will not encounter the Meridian flip issue.

32. So, if I accidently unplug power to my mount, do I have to start all over and do another alignment?

You can recover without having to totally start over. Turn the power switch off and plug your power back in. Unlock each axis and move them back to their index marks and lock them back down. Now turn the power back on to begin the alignment process and enter the current time (and date if necessary) and location as usual. When prompted for the alignment type, use the buttons to select Last Alignment and you're done! Now you can select the object you were viewing and slew back to it. You should be OK for general viewing. If you are doing astrophotography, you will probably need to slew to your alignment stars and press Align to fine tune their centering.

33. So, what is available to help me know what I can and can't see?

Have you ever selected an object using the hand controller and the telescope slews and points to a tree or something else that blocks your view? This is particularly frustrating during initial alignment or adding calibration stars if you have things around you that affect your effective viewing area.

If you have an iPhone or iPad, you can use a planetarium app like SkySafari, StarSeek 3, Starmap, Pocket Universe or SkyQ to look up the star or object before you select it on the hand controller to get a better idea of its current location in the sky and whether or not it might be blocked by something nearby. You can find similar apps for Android phones, including SkySafari.

I also highly recommend getting the app called Observer Pro if you have an iPhone and you want to find out what objects you can and can't see at any point in time. It comes with great visual tools to help you know how visible many popular objects will be, including effects of moon location and phase.

34. So, is there any way I can easily capture and use what my horizon actually looks like?

Wouldn't it be great if you could have your actual horizon in an application so you would know at any point in time what might be blocked out? The application Observer Pro allows you to use your iPhone or iPad camera to quickly capture an outline of your actual horizon!

Even though it is just a filled outline of your horizon, it has many uses. Observer Pro modifies its charts based on your captured horizon information. It is good for planning your evening or spot checking an object before you slew to it.

In this screen of a list of objects, you can see it shows there is a viewing gap for NGC 6218 (M12) for a portion of the night from my driveway.

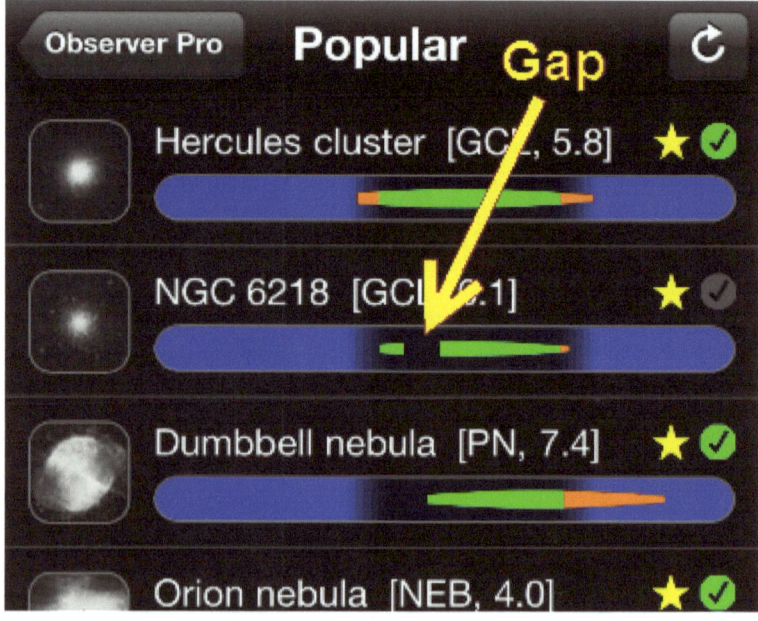

You can also select an object to better see its 24 Hour Visibility chart and see a sky view showing its path through your actual horizon.

Here you can see how the top of some trees in my backyard are the culprit that block its visibility for a portion of the night.

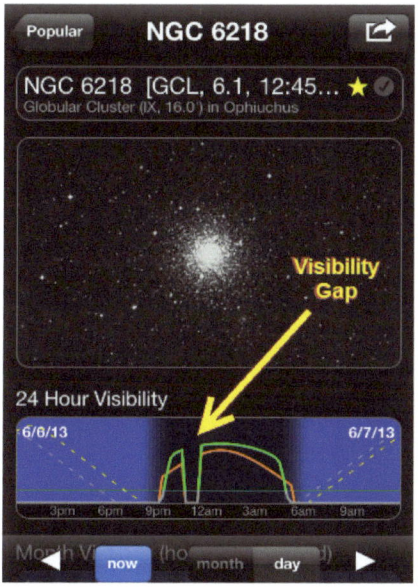

Guess what else you can do with this horizon data? If you are computer savvy, you can email your horizon file from Observer Pro to yourself, save and process it on a PC (with a script you download at Observer Pro's website: observerpro.com/converters) to convert your horizon file into a png file that can then be used in SkySafari. Here is a SkySafari screen showing the location of M12 next to my "tree horizon" using the image generated from Observer Pro.

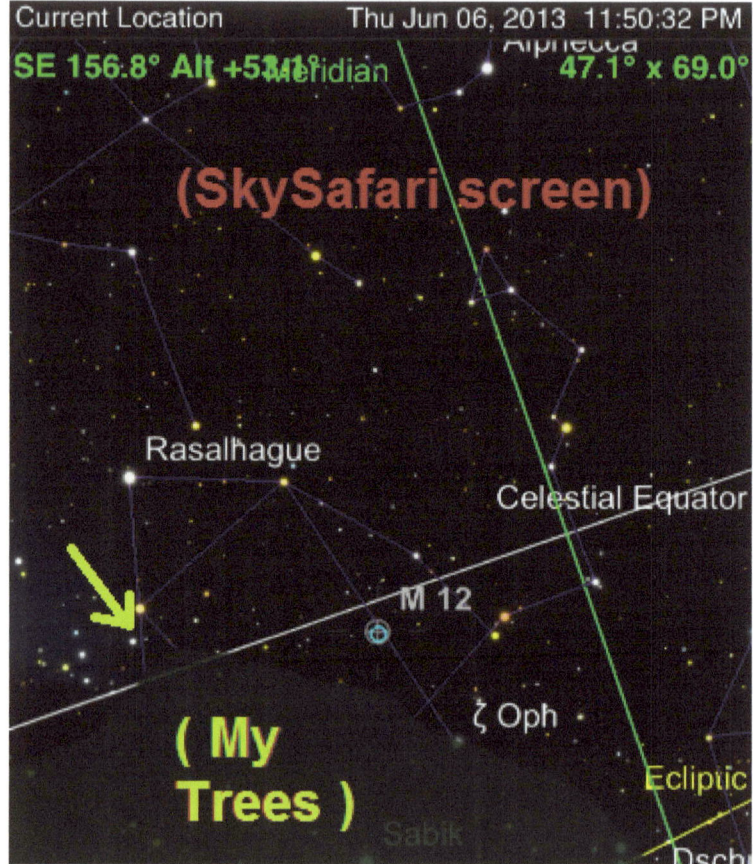

    I find this ability to look up objects in SkySafari with my actual horizon shown on my iPhone is a particularly valuable tool during alignment of an equatorial mount, which is why I have included this as a question. As of this printing, Observer Pro does not yet include individual stars in its database, but with my horizon inserted into SkySafari, I can check any star I want on the spot to see if it is visible at that point in time. This helps if you have any question whether the alignment star it selected is currently visible before it slews to it. It is very valuable when adding calibration stars since some of the stars the hand controller selects may not be as well known to you! If you determine a calibration star is currently blocked due to your limited horizon view, just press Undo on the hand controller to let it pick another one.

I have included in the appendix a description of the way you can also use this converted png file of your horizon in the planetary program Starry Night for use on your PC. This is very useful if you use Starry Night to control and slew your mount for you, since it is immediately obvious on the laptop screen whether you will be able to see an object before you issue the slew command.

35. So, can I use my Celestron alt-az AC power cord on my Celestron equatorial mount?

The answer is nope. Not enough juice. You may see slewing errors if you try to use the AC power adapter from a Celestron Alt-Az mount on your CG-5 mount because it sometimes requires a lot more current than the basic Alt-Az mount does. Instead, you can buy the Celestron CGEM AC adapter which provides the required amps and has the right type of mount connector that stays in place.

Some use an AC power supply like the Pyramid PS-9K 5 Amp Power Supply with Cigarette Lighter Socket to plug in the cigarette lighter end of the CG-5 power cable that came with your mount and run it off AC.

36. **So, can I use my Celestron alt-az hand controller on my Celestron equatorial mount?**

Here is another "it depends" answer. If something happens to the hand controller that comes with the CG-5 mount and you still have the hand controller from your Alt-Az mount, you can plug it into the CG-5 mount and move the mount using the directional arrows. However, don't try to align it or use the GoTo functions unless you reload the firmware in the Celestron Alt-Az hand controller with the GEM version of the firmware. This process requires a standard Hand Control RS-232 Port Cable to connect the hand controller to your computer. If your computer does not have a serial port you will also need a USB-to-serial adapter available at most stores that sell computers. You can find a good description of this process on the NexStar Resource Site at www.nexstarsite.com/OddsNEnds/HCFirmwareUpgradeHowToVersion4.htm . You can later reload the original alt-az firmware back into the hand controller if you want to use with your alt-az mount again.

37. **So, can I extend the hand controller for the mount with a long 6 conductor phone cord?**

This is not a good do-it-yourself thing to try unless you have the means to verify the connections before using it. Do not go down to Radio Shack and buy a 25 foot 6 conductor phone cable and adapter. It is too easy to get the connections reversed. If you are not careful, you can fry your hand controller. Only buy a cable that has specifically been made and tested for use with the Celestron Hand Controller, like the Handbox Extension Cable #CGMX for Celestron Mounts with NexStar Hand Controllers.

38. **So, are there other ways I can extend the control of the Celestron mount?**

There are a couple of ways you can effectively extend your control of the CG-5 mount using an iPhone or iPad. Orion's StarSeek Telescope Control Cable can connect to an iPhone/iPad to use with Orion's StarSeek astronomy apps. Southern Stars SkyWire cable can also add a serial port to your iPhone/iPad that is compatible with the standard Hand Control RS-232 Port Cable that connects to the bottom of the Celestron hand controller.

If you have an iPhone that uses the newer Lightning connector, you will need Apple's Lightning-to-30-pin adapter or cable. SkyWire allows your Celestron mount to be controlled by the iPhone/iPad SkySafari application. See southernstars.com/products/skywire/index.html. SkyWire is also compatible with other equatorial mounts as well.

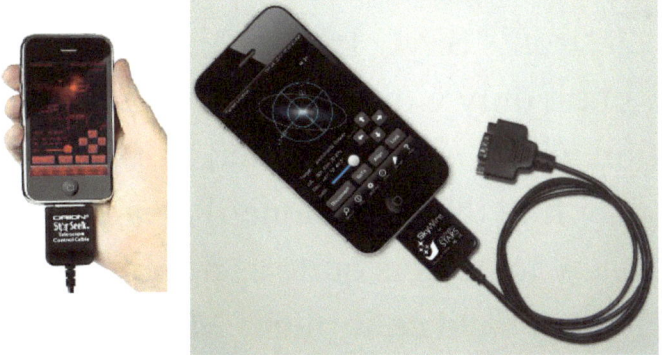

You can wirelessly extend your control of your Celestron mount with either the Southern Star SkyFi module or Orion's StarSeek WiFi

Telescope Control Module (Orion licenses it from Southern Star).

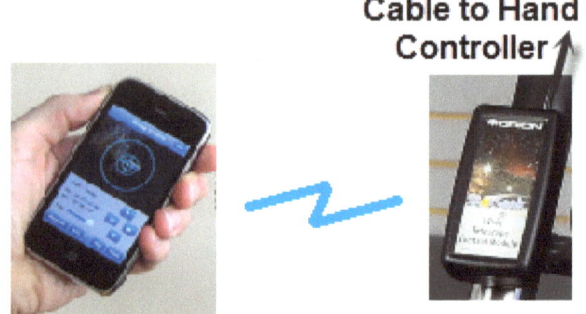

The wifi module connects to the Celestron hand controller's serial port and transmits over wifi to SkySafari, StarSeek 3 or Starmap Pro on the iPhone/iPad. You can use this in the field since they can create their own wifi connection directly with the iPhone or iPad. Skyfi also works with the SkySafari application on the Mac. Celestron's SkyQ Link is also a wifi module, but it only works with its own SkyQ application which is more limited.

You may find that an iPhone/iPad application does not slew to objects quite as accurately as your Celestron hand controller. Make sure your time and latitude/longitude location used are the same for both the hand controller and the app (your phone is a good source for accurate time and latitude/longitude). You can also use the application's align/sync function to improve its GoTo's. Find a bright star on the screen near your object of interest and use the application to slew to it. Then use the application's arrows or your hand controller arrow buttons to center the star in the eyepiece. Select the application's align/sync function which will then cause the application to re-center itself on that star. Now when you use the application's GoTo function it will more accurately center objects.

You can also control your telescope through your laptop's wifi by using a utility like Com2Tcp to create a virtual serial port on a PC that communicates to your mount via SkyFi. You can then control your mount wirelessly using the Celestron NexRemote virtual hand controller running on your laptop.

If you install ACOM drivers you can also wirelessly control your mount from a PC planetary application such as Starry Night.

39. So, is there anything simple I can do to streamline my setup process?

Setting up an equatorial mount and aligning it is just going to take longer than the setup process for the alt-az mount. However, there are some things you can do to streamline and reduce the time required for setup. I have a favorite spot on my driveway for my telescope. After performing a good polar align one evening, I marked the three spots on the driveway with black tape at the points where the tripod legs were located. I also used a black magic marker to mark around the lower legs to indicate where my tripod legs were extended. Then whenever I set up, I extend the legs to the same points and position the tripod at the taped marks.

This means the tripod is practically level and pointed very close the celestial pole, so you can set up during daylight and go ahead and attach your equipment while you can see better. You may still need to do a polar alignment, but I have found that sometimes it was close enough for basic visual viewing without having to do a polar align. This is true, especially if you left your mount attached to your tripod when you stored it since it will approximately retain the alt-az polar position you previously set.

Here is a case where all I did was position the tripod & mount at the driveway marks and performed a two star align and added one calibration star, and that was all I needed for my viewing session that night.

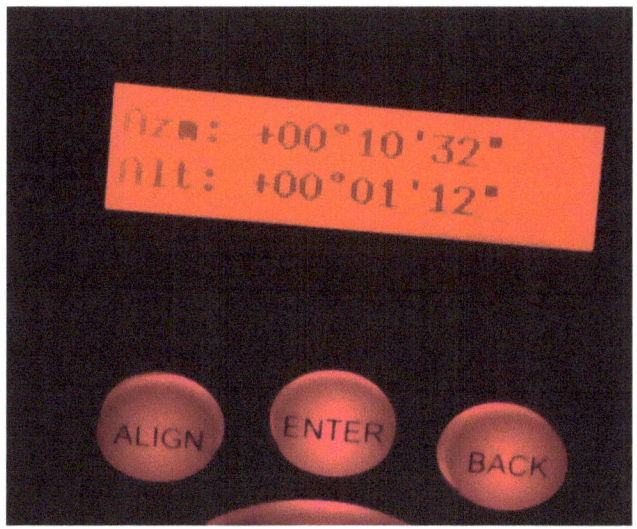

Of course an even better way to do this is install a pier in your back yard that stays level and is positioned well for the celestial pole, but I have not done that yet. Maybe some day.

The other thing you can do is to put your tripod on wheels so you leave more of your gear assembled and still easily move it from a covered location (i.e. garage) back into place. See question 40.

Adding Celestron's new StarSense accessory is another way you can reduce your alignment time. See question 48.

40. So, is there a good way to move the tripod, mount & telescope around when it is fully setup?

You can purchase a wheel assemble for your tripod so you can easily move it while it is set up. JMI's Medium Size Universal Wheely Bars allow you to move your tripod and mount already assembled.

However, I don't recommend trying to move it with the telescope attached if you have the standard small wheels unless your surface is absolutely smooth. However, if you purchase it with the large pneumatic wheels you will have a heavy duty method to easily move your fully assembled tripod, mount & telescope. This really speeds up your setup time if you have a place to store it, and just roll the assembly out into position to use it.

Another option is the ScopeBuggy which provides a medium duty mobile platform.

41. So, what are some things I should be careful to avoid?

When you initially set up your tripod and mount, do not put your counterweights on until you have the tripod positioned toward the celestial pole. Moving the tripod even a little with counterweights in place increases the risk of accidently tipping it over, and it is harder to adjust its position. Also, the less the legs are extended, the more easily it may tip over when the counterweights are attached.

If you are moving your tripod and mount assembly using small wheels, you should probably do it without the counterweights attached. If a wheel briefly sticks, the tripod could easily tip over. Depending upon your latitude, you may want to relocate your alignment metal peg for better stability. See question 42.

You should make sure your counterweights are not going to hit your mount as it moves. I always position them with their knob facing away from the tripod when the mount is at its index marks. For lower latitudes, you may want to remove the front altitude adjustment screw as described in the manual to avoid it coming into contact with the RA motor assembly as it moves.

Just use the rear adjustment screw lever for all your altitude adjustments.

Don't forget to put the safety screw back in place after you put your counterweight(s) on. When I remove the safety screw, I put it upside down on the eyepiece tray as a reminder, and so I can easily grab it when I am ready to put it back on.

42. So, is there any way to position the counterweights over a tripod leg for better stability?

Depending upon your latitude, you can relocate the alignment metal peg so the counterweights will be over a tripod leg for better stability.

There are two threaded holes in the tripod base where the azimuth adjustment peg can be installed. This is the metal peg that the azimuth housing of your mount fits over when attaching it to the base. This peg acts as a guide and a stop as you turn the two azimuth adjustment screws to polar align your mount. One of the holes located between the two tripod legs is typically where the peg is installed at the factory. Having it in this location positions the counterweight between the legs and allows the mount to be used at low latitudes without hitting the legs. However, the tripod can be less stable with the peg in this location, especially before everything is balanced. The other hole is over one of the tripod legs. Moving the metal peg to this hole results in a much more stable tripod setup.

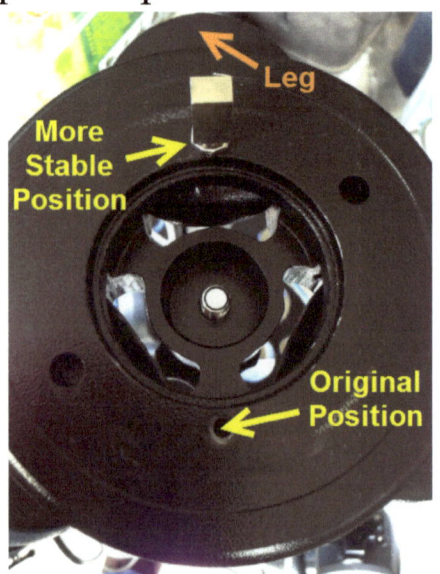

Use a small wrench to hold the peg in place while using another wrench to loosen the locking nut below the peg, and then unscrew the peg. Screw it into the hole over the tripod leg until it stops. Then back it out a little until it is square with the tripod. Retighten the nut with a wrench while holding the metal peg stationary with another wrench.

43. So, what do error messages "No Response 16" & "No Response 17" on the Celestron hand controller mean?

These messages indicate there is a communication loss between the hand controller and the motor controller chips for the DEC or RA axis. This typically occurs when there is some power issue with one of the axes while slewing the telescope. It could be simply that your battery is running down. Or, you could have a loose connector. Make sure the Dec cable is properly connected on both ends. If you continue to see these messages you may have developed a problem with your hand controller.

I have also seen this issue if I plan on controlling the mount from my PC and have the hand controller connected to my PC before I turn my mount on. To resolve this, turn the mount back off and disconnect the hand controller from the PC. Turn the mount back on and then reconnect it to the PC and see if that fixes the problem.

44. **So, is the mount OK if it is kinda noisy?**

Well, if you have the Celestron CG-5 it is just going to be somewhat noisy when it slews. This is a characteristic of the CG-5 mount. When it starts tracking, though, it should be pretty quiet. Some of the heavier duty equatorial mounts will be less noisy, but then you will pay more for them.

45. **So, what does hypertuning an equatorial mount mean?**

HyperTune is a process developed by Deep Space Products for taking a mount apart to change out some of the parts, clean it and use higher grade grease to improve the mount movements and accuracy. You can buy a kit to do it yourself if you are mechanically adept, or you can send in your mount and have it done for you. See www.deepspaceproducts.com/hypertune.html . They have two kits for the CG-5. The standard kit comes with a PTFE bearing set to replace the mount's plastic bearings, and the advanced kit provides a new complete worm gear ceramic hybrid ball bearing set for the mount. They state the process will take 2-3 hours, but you probably should count on longer and take plenty of time.

If your mount is performing well and doing everything you need it to do, then there may be little or no need to hypertune it. However, when a mount is not performing up to reasonable expectations, particularly in regards to balancing, backlash and tracking, then hypertuning can often be very helpful.

If you send in your mount to be serviced, they perform high resolution sanding and polishing of the internal moving parts to produce very smooth performance. I personally have not done this yet, so I cannot recommend this one way or another. But I have read posts by others saying it definitely improved the mount's performance, especially for astrophotography.

46. So, should I get Celestron's new Advanced VX mount instead of a CG-5?

As this book was being published, Celestron released a new Advanced VX mount very similar to the CG-5 with several improvements. Some of the new features include programmable Periodic Error Correction with integer gear ratios, new motors for improved tracking performance, viewing across the Meridian without interference from the motor housing and a number of redesigned mechanical features to make it easier to set up and use. It would be well worth paying a little bit more for the new model if you are able to do so.

However, the CG-5 is a great workhorse that has performed well for many over the years. You may be able to get a CG-5 at a good price as the Advanced VX gets into the market.

47. So, do I need to use PEC or autoguiding with an equatorial mount?

I know you now know what I'm going to say. It depends. PEC or autoguiding is mainly needed when you get into long exposure astrophotography. Autoguiding is especially useful when photographing deep sky objects. But for basic visual viewing sessions, as long as you perform a good alignment, you should be fine without using either of these. Performing a good Polar Align will even give you good tracking for trying out astrophotography. If you get serious about exposures over 2-3 minutes, you should consider adding an autoguider, which will monitor the position of a guide star you select and send commands to your mount to keep the guide star in the same visual position. This overcomes problems with tracking accuracy due to periodic errors in the gears and other sources of tracking errors.

Periodic Error Correction (PEC) is a method to correct the small mechanical error in the accuracy of the tracking that repeats at a regular interval - the interval being the amount of time it takes the mount's drive gear to complete one revolution. This small error in motion is often not noticeable during visual observation, but during a long exposure it will change stars from pinpoints to small streaks. The Advanced VX mount has the ability to record the periodic error over a 10 minute period, and then play back the corrections as needed. The CG-5 does not support PEC in its hand controller. However, when you use an autoguider on the CG-5 (or any GEM), it corrects for tracking errors which includes the periodic error from the worm drive.

There are different methods to use for autoguiding. An autoguider is usually attached to either a guidescope (a smaller telescope mounted on the main telescope) or to an off-axis guider, which diverts some of the light originally headed towards the eyepiece.

An autoguider has an image sensor that regularly makes short exposures of an area of sky near your object of interest. After each image is captured, a computer measures the apparent motion of one or more stars and issues the appropriate corrections to the telescope's computerized mount.

Star images will typically affect more than one pixel in the image. Autoguiders use the amount of light falling on each pixel to calculate where the star is to sub pixel accuracy and thus can track stars to an accuracy better than the angular size represented by one pixel. However, seeing conditions can limit its accuracy. To prevent the telescope from moving in response to changes in the guide star's apparent position caused by seeing conditions, you can adjust a setting typically called aggressiveness.

Autoguiding is a learning process involving selecting a good guide star and adjusting parameters for best results. I have found using the Orion Magnificent Mini AutoGuider Package (their Star Seek Auto Guider plus their 50mm guidescope) is a good lightweight assembly to get into autoguiding if you have a laptop you can use with it. It works well with PHD autoguiding software.

The Orion Star Seek AutoGuider (SSAG) can also be used with an off-axis guider placed in line with your camera. A minor portion of the light coming through your telescope is diverted by a small mirror to the autoguider.

With this arrangement you do not have to have a separately mounted guidescope thus eliminating any problems due to differential flexure (differences in flexure of the imaging camera and the autoguider camera). However, it does require setup time to get your main imaging camera and your autoguider focused properly depending upon the amount of difference in their back focus. It is a good idea to work with an off-axis guider first during daylight to learn how to best obtain focus with both cameras in place.

As I got into astrophotography and video astronomy, I discovered another use for the Orion Mini Autoguider Package. I actually now also use it as a remote finderscope. As I use software on a laptop to slew the telescope to objects and make final centering adjustments, I can watch the guidescope image on the screen to center my DSO before I even switch to using my main camera. The Orion SSAG with PHD software allows exposures up to 10 seconds that will show many DSO's position relative to surrounding stars (small and faint perhaps, but you can see it). You can then lower its exposure to 1-2 seconds and use those reference stars around the DSO area to center it for your main camera. The Mini Autoguider Package also uses the same quick release mounting bracket as the 9x50 finder scope described in question 28. Since the SSAG is removable, you can also insert other types of cameras with more sensitivity as remote finders as well (e.g. Mallicam Jr). The Orion 9x50 guide scope is also available by itself without the SSAG if you already have a camera you want to use instead.

48. So, how does Celestron's new StarSense help align a German Equatorial Mount?

I have a StarSense on order but did not receive it before publication, so the following information is based on the experience of others.

The StarSense accessory replaces the alignment process of finding and centering two alignment stars. StarSense automatically aligns itself with no other input needed and is compatible with all of Celestron's current computerized mounts/telescopes including the CG-5 and VX mounts. The StarSense accessory includes a small digital camera that attaches to the telescope's optical tube, along with a specialized hand controller.

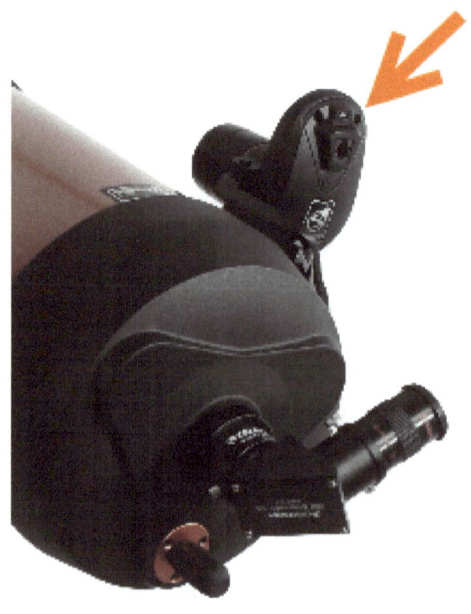

After setting up the StarSense accessory, you push the Align button on the hand controller, and StarSense starts capturing and comparing images of the night sky against its internal database of 40,000 celestial objects. In about three minutes, StarSense will gather enough information to triangulate its position and align itself. You can then use it to select and slew your telescope to the objects you want to view.

If your horizon is partially blocked, I have found that you can still use StarSense in a semi-manual mode to pick your visible portions of the sky for it to use for alignment. Move your telescope to a leftmost clear area of the sky for your first target area. Then move right to your next most clear area of the sky for the next target area. Then move right or up to the final clear area of sky for it to complete its alignment process. You do not have to find or center any objects – you just assist by pointing it to clear portions of the sky.

You will still need to initially set up your mount aimed at the celestial pole in your normal manner. After using StarSense to perform your alignment, you may want to do an All Star Polar

Align if you need additional alignment accuracy. You can also add as many calibration stars as desired to give you even higher level of accuracy.

Question List

So…

1. What is an equatorial mount and why is it different from an altitude-azimuth (alt-az) mount?
2. Give me the Top 10 reasons to switch to an equatorial mount.
3. Give me the Top 5 reasons to not switch to an equatorial mount.
4. Is a German Equatorial Mount better than other equatorial mounts?
5. What are the different kinds of equatorial mounts?
6. Can I convert my alt-az mount to an equatorial mount?
7. Could I just add an autoguider to my alt-az mount to get the same tracking accuracy as an equatorial mount?
8. What is involved in setting up a GEM?
9. How much time should I spend leveling my tripod?

10. How tight should I screw the tripod rod into the mount?
11. How much should I screw in the knob to tighten the eyepiece tray?
12. Are there any suggestions about balancing a GEM?
13. Why are there two screws to tighten the telescope onto the mount?
14. Will I need additional counterweights and, if so, what size?
15. I never used the setting circles markings on my alt-az mount, do I need to use them on an equatorial mount?
16. What are those circles with numbers under the removable cap for?
17. Why does the equatorial mount alignment process take longer?
18. What is done for basic alignment?
19. What additional steps can be added for better alignment?
20. What optional steps can be added for even better GoTo's and tracking?
21. What are the different methods available to perform a polar alignment beyond just pointing the mount North?

22. Do you really have to do a full polar alignment?
23. Once you do a polar alignment why do you have to do another standard alignment?
24. Why are calibration stars needed?
25. How do you mount a polar axis finderscope?
26. How do I use a polar axis finderscope?
27. Why does my view jump when I am adjusting the azimuth knobs during a polar alignment?
28. What sighting / finder scopes should I use?
29. Is there a special solar finderscope for pointing your telescope at the sun?
30. How do I aim this thing?
31. What is meridian flip and can it break anything?
32. If I accidently unplug power to my mount, do I have to start over and do another alignment?
33. What is available to help me know what I can and can't see?

34. Is there any way I can easily capture and use what my horizon actually looks like?
35. Can I use my Celestron alt-az AC power cord on my Celestron equatorial mount?
36. Can I use my Celestron alt-az hand controller on my Celestron equatorial mount?
37. Can I extend the hand controller for the mount with a long 6 conductor phone cord?
38. Are there other ways I can extend the control of the Celestron mount?
39. Is there anything simple I can do to streamline my setup process?
40. Is there a good way to move the tripod, mount & telescope around when it is fully set up?
41. What are some things I should be careful to avoid?
42. Is there any way to position the counterweights over a tripod leg for better stability?

43. What do error messages "No Response 16" & "No Response 17" on the Celestron hand controller mean?
44. Is the mount OK if it is kinda noisy?
45. What does hypertuning an equatorial mount mean?
46. Should I get Celestron's new Advanced VX mount instead of a CG-5?
47. Do I need to use PEC or autoguiding with an equatorial mount?
48. How does Celestron's new StarSense help align a German Equatorial Mount?

Appendix

Below is a list of steps that tell how to set up horizon data from Observer Pro in Starry Night Pro 6. This assumes you have extracted a hzn file from Observer Pro and converted it to a file called MyPatio.png as discussed in question 33.

1. Go to the "...\Program Files\Starry Night Pro 6\Sky Data\Horizon Panoramas" folder.
2. Copy MyPatio.png created from your hzn file into this folder.
3. Make a copy of 01_Earth_Lake.txt and rename it 00_MyPatio.txt (Using 00_ will put it at the top of the list)
4. Change the following values in 00_MyPatio.txt* in order to use MyPatio.png created from the hzn data:

"PanoName" value="MyPatio"

"ImageFileName" value="MyPatio.png"

"UseImageAlpha" value="Yes"
"ImageHeight" value="179.900000000000000000"

"ImageCenterDec" value="0.000000000000000000"

"ImageCenterRa" value="180.000000000000000000"

5. Save your changes.
6. Start up Starry Night and click on File, Set Home Location, Select Panaroma.
7. Click on the MyPatio and then on Select and then Save as Home Location.
8. Your Observer Pro horizon will now be overlaid on the Starry Night sky!

*The changes made to MyPatio.txt are highlighted below:

```
<HTML><BODY>This file is a <A HREF="http://www.starrynight.com">Starry Night
document.</A><SN_VALUE name="Version" value="Pro (Mac) - p400c-EM"><SN_VALUE
name="VersionSKU" value="p400c-EM"><SN_VALUE name="charset" value="Macintosh-
ASCII">// Pano name - you see this in menus, etc. This should be unique for each
pano<SN_VALUE name="PanoName" value="MyPatio"><SN_VALUE name="ImageFileName"
value="MyPatio.png">// In degrees, the height of the image<SN_VALUE
name="ImageHeight" value="179.90000000000000000">// If the image has an alpha
channel that is set up with the sky having an alpha of 0 and the horizon an
alpha of 1, // with alpha blending at the interface, we need to say Yes here,
otherwise say "no"<SN_VALUE name="UseImageAlpha" value="Yes">// These values are
in the coord system that you pick out with ImageCoordSys<SN_VALUE
name="ImageCentreDec" value="0.000000000000000000"><SN_VALUE
name="ImageCentreRa" value="180.000000000000000000">// These are also in
ImageCoordSys coord sys// They determine where the image will be <SN_VALUE
name="ImageAxisDec" value="90.000000000000000000"><SN_VALUE name="ImageAxisRa"
value="0.000000000000000000"><SN_VALUE name="PanoBrightness"
value="1.000000000000000000">// Used during liftoff to draw a horizon nicely.
Ignored for panos like the milkyway which are always far off.<SN_VALUE
name="PanoBottomDistance" value="0.1"><SN_VALUE name="PanoTopDistance"
value="1.000000000000000000">// Used during liftoff to draw a horizon nicely.
Ignored for panos like the milkyway which are always far off.// the color is
red, green, blue, with each value scaled 0 to 65535<SN_VALUE
name="PanoApproxColor" value="36000, 36000, 36000">// If preload is set, will
load image on startup<SN_VALUE name="Preload" value="Yes">// ImageCoordSys
values//                              kNoCoordinateSystem = 0, //
        kAltAzSystem = 1, //                         kGalacticSystem = 2, //
                kEclipticSystem = 3, //
kCelestialJ2000System = 4, //                          kCelestialJNowSystem =
5, //                       kOrientationSystem = 6, //
kSuperGalacticSystem = 7,<SN_VALUE name="ImageCoordSys" value="1">//
ImageTransferMode values// OpenGLState  kNoState = 0,//
kDefault = 1,//                         kStars = 2,//
kTransparentAdd = 3,//                            kTransparentGlass = 4,//
                kOpaque = 5,//                           kShaded = 6,//
        kLighting = 7,        // shaded with specular lighting//
        kShadedLine = 8,//                       kBillboard = 9,// For
horizons, 4 (kTransparentGlass) and 5 (kOpaque) are <SN_VALUE
name="ImageTransferMode" value="4"><SN_VALUE name="ImageWidth"
value="360.000000000000000000"></BODY>
```

Here is what MyPatio.png looks like:

Here is Starry Night using MyPatio.png:

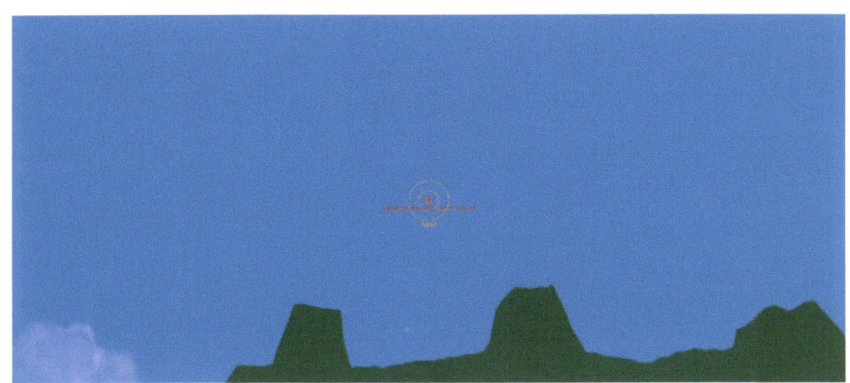

Suggestions

Send suggestions, corrections or other feedback to:

BeginSoWTM@aol.com

www.ingramcontent.com/pod-product-compliance
Lightning Source LLC
Chambersburg PA
CBHW040825180526

45159CB00001B/72